SCIENTIFIC AMERICAN EDUCATIONAL PUBLISHING

# DO THE MATH!

## 10 FUN PROJECTS USING MATH

BRING SCIENCE HOME

Published in 2024 by Scientific American Educational Publishing in association with
The Rosen Publishing Group 29 East 21st Street, New York, NY 10010

Contains material from Scientific American, a division of Springer Nature America, Inc., reprinted by permission as well as original material from The Rosen Publishing Group .



First Edition

Activity on page 5 by Science Buddies (December 3, 2015); page 11 by Science Buddies (September 24, 2015); page 17 by Science Buddies (January 8, 2015); page 23 by Science Buddies (November 29, 2012); page 29 by Science Buddies (June 6, 2013); page 35 by Science Buddies (September 27, 2012); page 41 by Science Buddies (April 5, 2012); page 47 by Science Buddies (March 29, 2012); page 51 by Science Buddies (March 29, 2012); page 55 by Science Buddies (March 15, 2012).



Editor: Kelly Meredith
Designer: Rachel Rising

All illustrations by Continuum Content Solutions

Photo Credits: pp. 3, 5, 8, 14, 15, 20, 23, 27, 29, 32, 35, 38, 44, 45, 49, 53, 54, 58, 59 cve iv/Shutterstock.com; pp. 4, 5, 11, 17, 23, 29, 35, 41, 47, 51, 55 Anna Frajtova/Shutterstock.com.

Cataloging-in-Publication Data

Names: Scientific American, Inc.
Title: Do the math!: 10 projects using math / edited by the Scientific American Editors.
Description: New York : Scientific American Educational Publishing, 2024. | Series: Bring science home | Includes glossary and index.
Identifiers: ISBN 9781725349032 (pbk.) | ISBN 9781725349049 (library bound) | ISBN 9781725349056 (ebook)
Subjects: LCSH: Mathematics--Experiments--Juvenile literature. | Science projects--Juvenile literature.
Classification: LCC QA40.5 D64 2024 | DDC 510--dc23

Manufactured in the United States of America

Some of the images in this book illustrate individuals who are models. The depictions do not imply actual situations or events.

CPSIA Compliance Information: Batch #SACS24. For further information contact Rosen Publishing, New York, New York at 1-800-237-9932.

# CONTENTS



* THESE ACTIVITIES INCLUDE SCIENCE FAIR PROJECT IDEAS.

# INTRODUCTION

Math is a tool we use every day, even in places we don't expect it! In these experiments, you will use candy to help predict outcomes, Play-Doh to investigate changes in shape volume, and playing cards to figure out odds—all with math at the center.

Projects marked with include a section called Science Fair Project ideas. These ideas can help you develop your own original science fair project. Science fair judges tend to reward creative thought and imagination, and it helps if you are really interested in your project. You will also need to follow the scientific method. See page 61 for more information about that.

# Half-Life Coins

DID YOU KNOW THAT RADIOACTIVE MATERIAL DECAYS RANDOMLY? LEARN HOW FLIPPING COINS CAN HELP YOU FIGURE OUT HOW LONG RADIOACTIVE MATERIAL STAYS DANGEROUS IN THIS FUN (AND PERFECTLY SAFE) ACTIVITY!

## PROJECT TIME

1 to 2 Hours

## KEY CONCEPTS

Energy
Radioactivity
Exponential decay
Odds

One way of creating energy is with nuclear reactors. These plants are generally safe, but occasionally there are accidents in which dangerous radioactive material escapes. You might have read about nuclear disasters, such as those at Chernobyl, Three Mile Island, and Fukushima, in the news or in a history lesson. Disasters like these can take years or even decades to clean up and make it unsafe for humans to live nearby for even longer. Why does the contamination last so long? In this (perfectly safe!) activity, you will learn about radioactive decay using coins.

## BACKGROUND

All matter is made of atoms. Certain types of atoms are "radioactive," meaning that they will eventually decay, or "break down" into a different type of atom. When this breakdown process occurs, the atom emits radiation. Some types of radiation surround us every day and are perfectly safe, such as radio waves and visible light; other types, such as X-rays and gamma rays, can be dangerous to humans. The types of atoms used as fuel in nuclear reactors can produce such dangerous radiation, which is why it is very important to keep the fuels safely contained. Thus it can be very dangerous when there is an explosion—or in the case of Fukushima, an earthquake—at a nuclear plant, and some of the radioactive atoms escape into the surrounding air, water, or soil, causing contamination.

The decay of radioactive materials is a random process, kind of like flipping a coin or rolling a die. At any given moment in time, there is a chance that an atom will decay, but there is also a chance it will remain the same. The rate at which radioactive materials decay is measured with something called the "half-life." The half-life describes how long, on average, it takes until one-half of the original radioactive atoms are left. The half-lives of different atoms can vary widely—some are less than a second, and others are thousands or even millions of years. In this activity, you will simulate radioactive decay by flipping coins. Coins that land tails-up "decay," and coins that land heads-up remain the same. This will allow you to learn more about the process of radioactive decay—without using any radioactive materials!

## MATERIALS

- About 120 coins (this can be a mix of pennies, nickels, dimes, and others)
- Resealable plastic bag
- Flat table top for counting coins
- Paper or notepad
- Pencil

## PREPARATION

- Assemble all of your materials at your workspace.
- Put exactly 100 coins in the resealable bag.

## PROCEDURE

- Shake the bag vigorously for a few seconds.
- Open the bag and carefully dump the coins out on a tabletop. Be careful not to dump them out too quickly, or they might roll onto the floor.
- Separate the coins into two piles: heads and tails. *About how many coins landed heads up, and how many landed tails up? What is the ratio between the two?*
- Count the number of heads. Write this number down.
- Put the coins that landed tails up aside. Put the coins that landed heads up back in the bag.
- Shake the bag again and repeat the process. Dump the coins out, separate the coins into heads and tails, count the number of heads, write the number down, and put the heads back in the bag. *About what ratio of heads/tails do you get each time?*
- Keep repeating this process until there are no more heads to put back in the bag, and you have set aside all 100 coins.
- Now, look at the numbers you wrote down. Make a stack of coins the same height as each number, and line the stacks up next to each other in order (this way you are making a "graph" using stacks of coins, instead of drawing one on paper). Note that you may need to use some of your extra coins to make the stacks (for example, say you wrote down the numbers 56, 25, 13, 6, 3, and 1. That adds up to 104 coins, so even though you only used 100 coins for the experiment, you need 104 to make the graph because some coins continued to be in the "heads-up" pile numerous times). *What does your "graph" look like? Does it form a straight line, or does it have a different shape? How many turns did it take you until there were no coins left?*

## EXTRA

Repeat the activity a few more times. *Do you get exactly the same results each time?*

## EXTRA

Repeat the activity with different numbers of coins. *What happens if you start out with 200 coins, or only 50 coins? Does the resulting graph have the same shape, or is it different?*

## EXTRA

Use a stopwatch to time how long you shake the bag to simulate materials with different half-lives, then use a ruler to space out your stacks of coins to represent time when you make your graph. For example, if you shake the bag for 10 seconds, you could place the stacks of coins 10 centimeters apart. *How does this graph compare to one where you shake the bag for 20 seconds, if in both cases you start out with 100 coins? What does this mean about materials with a longer half-life?*

## OBSERVATIONS AND RESULTS

You should have seen that the number of coins in the bag decreases by roughly, but not exactly, half each time you count heads and tails. For example, even if you always start out with 100 coins, that does not mean you will have exactly 50 heads and 50 tails the first time you shake the bag. You might get 56 heads and 44 or tails, or 49 heads and 51 tails. You will lose about half the coins each time, and it will probably take you about 6 turns until there are no coins left when you start out with 100 (remember that flipping a coin is a random process—so your results will not be exactly the same each time). The resulting graph does not form a straight line. Instead, it forms an exponential curve that starts off very steep, but then gradually tapers off towards zero. If you start out with more or fewer coins, the number of turns it takes you to reach zero coins will change, but the shape of the curve will remain the same.

This is why radioactive contamination is dangerous in the environment and hard to remove. Even though large amounts of the material may decay very quickly at first, smaller amounts can linger for a long time. For example, some of the contaminants in the Chernobyl disaster have half-lives of about 30 years. Imagine that you could re-do this experiment and wait 30 years until you repeated each turn. There would probably still be some coins left after more than 100 years! That is why, even though the Chernobyl disaster occurred in 1986, the area is still unsafe for humans to inhabit.

## CLEANUP

Return all materials to where you found them.

# Circular Reasoning
## Finding Pi

CAN YOU UNCOVER A USEFUL MATHEMATICAL FORMULA WITH JUST A FEW HOUSEHOLD OBJECTS AND SOME RIBBON? IT'S EASIER THAN YOU THINK. IN FACT, WE WOULD WAGER A SLICE OF PI THAT YOU CAN!

### PROJECT TIME

20 to 30 minutes

### KEY CONCEPTS

Circles
Circumference
Diameter
Mathematical formula
Pi

Mathematicians get excited about the discovery of mathematical relationships. They look at the world around them in terms of numbers, formulas, and equations. Mathematics is fun and practical too. It comes in handy when calculating how long you need to save your allowance before you can buy a new game. You use math to double or triple a recipe or to calculate how late you can leave your house and still get to school on time.

You can also use it to predict the size of things—on a page or in the real world. In this science activity, you will examine circular objects and see what relationships you can discover about their sizes. You will explore whether the circumference of a circle relates in a particular way to its diameter. As you explore the relationship, you might be surprised at how useful the results can be; maybe it will inspire you to save up for a bigger bike!

## BACKGROUND

Did you know people have been studying mathematics for thousands of years? Many mathematical terms we use today originate from Greek and Latin, revealing the origin of some of the ancient scholars who worked on these problems. The word circumference, for example, comes from the Latin words *circum* (which means around) and *ferre* (to carry). It can refer to the line bordering the circle as well as to the length of that line. The word diameter is derived from the Greek words *dia* (across) and *metron* (measure). It refers to a straight line that starts at a point on the circumference, goes through the center of the circle and ends at the other side of the circle. It can also refer to the length of this line. Mathematicians have proved that the diameter is also the longest distance across the circle.

Enough terminology! It's time to start exploring.

## MATERIALS

- At least four circular objects of different sizes (For instance, you can use a large coin, round container lid, large container lid, and a bicycle wheel.)
- A large roll of twine or ribbon that you can cut into small pieces
- Scissors (An adult should help you use them.)
- Tape, such as masking tape (optional)

## PREPARATION

- Assemble all of your objects in one area so that they are in easy reach.

- Start your exploration with a medium-size circle, such as a yogurt container lid. In the next couple of steps, you will cut pieces of twine (or ribbon) that have the length of the circumference and the diameter of this circle. Once you have those pieces ready, you can start exploring if these relate in a particular way. You will repeat the procedure for different-size circles in the hope of discovering that the diameter and circumference of all your examples relate to each other in the same way.

## PROCEDURE

- To create a piece of twine the length of the circumference (the line bordering the circle) of your first circular object, hold the end of a piece of twine, with your thumb, on a point on the edge of the circular object.

- Wrap the twine exactly one time around the object and cut the twine where the wrapped-around twine meets its starting point. To make this a little easier, you can temporarily attach the beginning of the twine to the circular object with tape, then wrap and cut the twine to that piece of tape. *Can you see that the length of your piece of twine is exactly the length of the circumference of your circular object?*

- To measure the diameter you need the length of a straight line that starts at a point on the circumference, goes through the center of the circle and ends at its other side. Because it is not easy to find a circle's center, you will use a mathematical fact about circles, which states that the diameter is also the longest distance across a circle. To create a piece of twine with the longest length across the circle use your thumb to hold the end of a new piece of twine (or ribbon) on a point on the edge of the medium-size circular object.

- Span a straight line of this twine across the circle to another point on the circumference of the circle. Now move the second point along the circumference—to the left and right. Do this until you find the longest straight spanned piece of twine possible. When you move the end of the twine away from this point, the spanned piece of twine gets shorter again. Cut off the piece of twine where it was longest to get a measure of the diameter of this circle. *Did you observe that your spanned twine went through the center of the circle?*

- Now you have everything you need to start exploring. *Which distance is longer—the diameter or the circumference? Is it longer by a lot or a little?*

- *If you fold the longer piece of twine in half, does it fit the length of the other piece?* If so, this would mean the longer piece is twice as long as the shorter piece. *If you did not find a good fit by folding the longer piece in two, what about if you fold it in threes, fours, or fives? Do you get an exact or an approximate fit? How would you translate your findings in words like "twice as long" or "three times as long"?*

- Try the activity again with a different-size circular object. *Do you expect the same relationship between circumference and diameter length to be valid for the different-size circle?*

- Repeat the circumference- and diameter-finding until you have explored a tiny, medium, big, and very large circle. *Can you find a relationship that is valid for all the tested circles? Is it an exact or an approximate relationship? If you found a relationship, do you think you have enough data to conclude that your relationship is valid for all circles?*

## EXTRA

Look around the house to locate some circular objects and make an estimate of the length of the diameter and the circumference of these objects. *Which one was easier to estimate for you, the diameter or the circumference? Which one would be easier to measure with a ruler?*

## EXTRA

*If you found an approximate relationship between circumference and diameter, how could you make it more exact?* Hint: you can use a ruler to measure the length of your twine pieces and do a little math. (For example, try dividing a circumference by its corresponding diameter; try again with each circle. *Do you get a similar number each time—regardless of whether the circle size was the same?)*

## EXTRA

One real-world application of this principle is when calculating the distance different-size wheels travel. To explore the relationship between the distance traveled on the ground and wheel size, mark a spot on the circumference of a wheel (such as a bicycle wheel) with tape. Place that spot on the ground and indicate this location on the ground with tape or chalk. Roll the wheel along a straight line until the same spot on the circumference touches the ground again. Mark this location on the ground with tape or chalk. Now compare the distance between the two marked locations on the ground with the length of the diameter and circumference of the wheel. *Can you find a relationship? What does this mean for the number of rotations a bigger or smaller wheel might need to travel the same distance?*

## EXTRA

*Now that you know the relationship of the diameter to the circumference—and maybe even the distance traveled—can you see practical ways where this relationship can be useful?*

## OBSERVATIONS AND RESULTS

If all went well, you should have discovered that the circumference was a little more than three times the diameter of the circle for each circle, no matter how small or big the circle was.

If you were able to work more exactly, you might have found that it was not exactly three times, but rather three and one-seventh the diameter. And even that is not exact.

Mathematicians found that the ratio of the circumference to the diameter of a circle is a constant, meaning it is the same for all circles, no matter how large or small the circles are. They also discovered, however, that this ratio is a number that can never be determined precisely. Since the mid-1800s, this ratio has been referred to with the Greek letter π (pi), which is a remarkably interesting number. It appears not only in geometry but also in other mathematics such as probability theory. It also shows up in the natural world, such as in the description of waves—from the visible ripples on the water to the invisible waves of light and sound.

### CLEANUP

Throw away cut pieces of twine or ribbon. Return all other items to where you found them.

# Shape Science

## Play-Doh Math

CAN SQUISHING PLAY-DOH HELP YOU LEARN ABOUT CHANGES IN AN OBJECT'S VOLUME?

Have you ever had fun making different figures or colorful shapes out of Play-Doh? You can squish and stretch a single piece of Play-Doh to make all sorts of shapes. How does changing the shape affect its volume? In this science activity you will find out by investigating how a piece of clay's shape affects its dimensions (length, width, and height) as well as how these changes are related to its volume.

### PROJECT TIME

45 to 60 minutes

### KEY CONCEPTS

Mathematics
Geometry
Three-dimensional
Volume

## BACKGROUND

Geometry is the study of how to use math to describe and investigate different points, lines, and shapes. A very basic three-dimensional shape is the rectangular prism, which is the shape of a box or a book; it has six sides. If all six sides are the same size and perfectly square, such as with dice, it is a cube. Areas and volumes of cubes and rectangular prisms can be measured with the same geometrical formulas.

Mathematical formulas describe geometrical shapes and are a way to calculate different properties of an object, such as its area and volume. Volume is a unique property of three-dimensional shapes because these shapes take up space in three different directions: length, width, and height. In this science activity, you will use Play-Doh (or homemade clay) to make a model of a rectangular prism, change one of its dimensions (height), and see what effect this has on the other two dimensions as well as the object's volume.

## MATERIALS

- Play-Doh (Alternatively, you can prepare homemade salt dough using plain flour, table salt, water, and other optional ingredients; see the "Preparation" section for details. Optional ingredients include vegetable oil, wallpaper paste, and lemon juice.)
- Permanent marker
- Ruler
- A table or desk
- A flat surface, such as a hard binder or book
- A scrap piece of paper and a pen or pencil

## PREPARATION

- If you want to make homemade salt dough, mix together 2 cups of plain flour, 1 cup of table salt, and 1 cup (237 ml) of water. Optional ingredients include 1 tablespoon (15 ml) of vegetable oil (to make it a little easier to knead), 1 tablespoon of wallpaper paste (to give the mixture more elasticity), and/or 1 tablespoon (15 ml) of lemon juice (to make the finished product harder).

## PROCEDURE

- Remove a chunk of dough about as large as your fist. Make your dough into a cube shape, approximately square on all sides.

- Using a permanent marker, label each of the three dimensions of the cube with a "W" (for "width"), "L" (for "length") and "H" (for "height"). On either side of each letter, put a dash to show the direction of the dimension on that side of the cube.

- Place your labeled cube of dough on a table or desk with the side marked for height pointing up and down (the dashes should be vertical). Use a ruler to measure the three dimensions you labeled on the cube (length, width, and height). *What are the measurements of the cube?* Write these down on a sheet of paper.

- Now you are ready to change the shape of your dough by squishing it. Put a flat surface, such as a book or a hard binder, on top of the dough cube. Slowly press down on the dough while keeping the corners square (straight) as you go by patting in from the sides with your hands. Stop pressing down when it looks like the dough has changed shape a little. *How does the dough look now? What shape is it?*

- Again, use a ruler to measure the three dimensions you labeled on the dough. *What are the measurements of the dough now?* Write these down.

- Again, squish the dough as you did before, trying to keep the corners square, and stop pressing down when the dough has visibly changed shape again. *How does the dough look this time?*
- Use a ruler one more time to measure the three dimensions you labeled on the dough. *What are the measurements of the dough this last time?* Write these down.
- For each time you measured the dough, multiply its length by width by height to calculate the volume of the shape in centimeters cubed. *What was the volume of the shape each time you squished the dough?*
- *Overall, what happens as one dimension (the one you flattened) decreases? Do the other two dimensions increase or decrease? Does changing the dimensions of the dough affect its volume or does the volume stay the same?*

## EXTRA

In this activity, you changed the length, width, and height of a cube of dough but you did not change the amount of dough you used. *What would happen if the amount of dough did change? Would the volume also change?* Try this activity again but this time add or take away some dough and form the dough into a cube, and measure the dimensions of the shape.

## EXTRA

You could take your results from this activity and graph them. A good type of graph to use for this activity is a bar graph. *Looking at your graph, how do the dimensions and volume change as the shape of the dough changes?*

## OBSERVATIONS AND RESULTS

As you squished the cube of dough, did its height get smaller, its width and length get larger, and its volume stay the same?

In this activity, because the amount of dough used did not change (you did not add dough or take any away), the size, or volume, of the dough should have stayed the same. Because the mathematical formula for calculating the volume of a cube is length X width X height and the volume stayed the same, as you squished the cube its height should have decreased (the cube was being flattened) whereas its length and width increased. For example, the initial cube may have had a length, width, and height each equal to 1.5 inches (4 cm). As the cube became squished, this may have changed to a length of 2 inches (5 cm), a width of 2 inches (5 cm) and a height of 1 inch (2.5 cm). Although the shape of the dough changed, its volume remained the same (each of these shapes have a volume of approximately 4 cubic inches, or 65 cubic centimeters).

## CLEANUP

Store Play-Doh or homemade dough in an airtight container to use again. Return all other materials to where you got them.

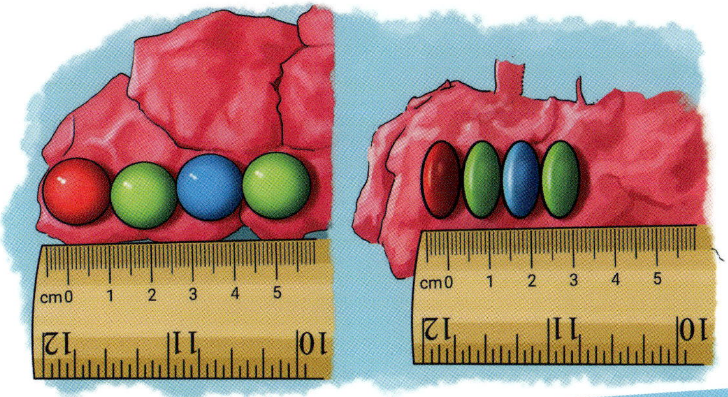


# Sizing Science

## The Geometry of M&Ms

CAN PLAY-DOH HELP YOU DETERMINE HOW MUCH CANDY YOU'VE EATEN? CHECK OUT THIS ACTIVITY TO SEE HOW THIS, ALONG WITH A FEW FORMULAS, CAN!

Has an adult ever caught you munching on candy and asked, "How much candy have you eaten?" Instead of saying, "I don't know," and possibly receiving a scolding, wouldn't you rather respond, "I ate precisely 10.7 cubic centimeters of candy"? In this activity, you will investigate which mathematical formula is most accurate for estimating the volume of an M&M. Figure this out and the next time you are discovered while snacking on sweets, you might make a better impression.

### PROJECT TIME

60 to 90 minutes

### KEY CONCEPTS

Geometry
Mathematics
Shapes
Dimensions
Volume

## BACKGROUND

Geometry uses math to describe and investigate different points, lines, and shapes. A shape is described in geometry using a formula, which is simply a mathematical way to calculate different properties, such as its size, area, or volume. Volume is a property of three-dimensional shapes—such as cubes and spheres—and takes into account the space the shapes take up in each of the three different directions.

The challenge of using geometric formulas in the real world is that these mathematical formulas often describe "perfect" or "ideal" shapes. A sphere is an "ideal" three-dimensional shape that is perfectly circular in all three directions. Even though a tennis ball, for example, is spherical in shape, it is not a perfect sphere (think of the lines that mark its surface). Most real-world objects are not simple shapes and require complex geometry to be calculated. The properties of real-world shapes can, however, be approximated, or estimated with a geometric formula. This is called "making a geometric model." The most important part of making a good geometric model is choosing the formula that best describes the object. Using that formula, you can geometrically model all kinds of irregular objects, such as cars, airplanes, toys, and M&Ms.

## MATERIALS

- Two pieces of paper
- Table or countertop
- Clay or Play-Doh. Use a small amount that you do not mind ruining.
- Metric measuring glass with milliliter markings
- Water
- 110 M&Ms. One 7-ounce (198-gram) bag holds about 210 M&Ms.
- Ruler with centimeter markings
- A pen or pencil
- A computer with an internet connection

## PREPARATION

- Place a sheet of paper on a clean table or countertop.

- On top of the paper, place a small amount of clay or Play-Doh. Flatten it and stretch it out into a line about 20 centimeters long by 3 centimeters wide. The line should be only a couple centimeters high. You will be measuring the M&Ms on this.
- The clay or Play-Doh may be ruined by the M&Ms' dyed coating, so use a portion that you will not mind ruining.
- Fill the metric measuring glass with 100 milliliters of water. If you intend to eat some of the wet M&Ms after this activity is completed, remember to clean the measuring glass carefully before you fill it with water.

## PROCEDURE

- Line up 10 M&Ms on their flat side, end to end, on the clay. Make sure the line is straight and flat and that each M&M is touching the next, with no gaps in between. You can poke them into the clay to keep them in a neat row.
- Measure the line of M&Ms in centimeters and divide this number by 10. This gives you the long diameter of a single M&M candy. *What is the long diameter of an M&M candy?* Write this measurement down.
- Divide the long diameter by two. This gives you the long radius of a single M&M candy. *What is the long radius of an M&M candy?* Write this down.
- Remove the M&Ms from the clay.
- Now line up the 10 M&Ms on their side so that you are measuring across the short side. Again, use the clay to hold them in a neat row in which each M&M touches the next.
- Measure the line of M&Ms and divide this number by 10. This gives you the short diameter of a single M&M candy. *What is the short diameter of an M&M candy?* Write this down.

- Divide the short diameter by two. This gives you the short radius of a single M&M candy. *What is the short radius of an M&M candy?* Write this down.

- Now you are ready to measure the actual volume of the M&M with a water-displacement test. Make sure that the metric measuring glass has exactly 100 milliliters of water in it (check by looking at the bottom of the meniscus).

- Dump 100 M&Ms into the glass of water. (If you want to eat soggy M&Ms later, use fresh M&Ms during this stage since the ones you used previously may have clay on them.) *What is the new water level of the glass?* Write this down.

- Subtract 100 milliliters from the new water level. Divide this number by 100. This is the actual volume of a single M&M in milliliters (which is the same as cubic centimeters). *What is the volume of one M&M?* Write this down.

- You will now be making some volume calculations using different formulas to see which one best calculates the volume of an M&M. Go to the Unit-Free Volume Calculators page at Mississippi State University: https://sdt13.abe.msstate.edu/tools/Volume/.

- Click on the "Full Sphere" link. For the "Radius" box, enter the long radius you measured for a single M&M and click "Calculate." *What is the M&M volume using its long radius?* Write this down.

- Now for the "Radius" box, enter the short radius instead. *What is the M&M volume using its short radius?* Write this down.

- Go back to the Unit-Free Volume Calculators page and click on the "Cylinder" link. For the "Outer Radius" box, enter the long radius; for the "Inner Radius" box, enter zero; and for the "Height" box, enter the short diameter. *What is the M&M volume using the cylinder formula?* Write this down.

- Go back and click on the "Ellipsoid" link. For the "Major Axis" and "Minor Axis" boxes, enter the long diameter (for both). And for the "Vertical Axis" box, enter the short diameter. *What is the M&M volume using the ellipsoid formula?*
- *How do each of the different calculated volumes compare to the actual volume that you measured? Which ones were more and which ones were less? Which calculation came the closest? Which formula do you think is the best one to use for an M&M candy?*

## EXTRA

Another way to look at your data is to calculate the percent difference between each calculation and the actual volume measurement. You can do this by dividing your answer using each formula by the actual volume. *Which formulas give you an answer that is closest to, or most different from, the actual volume based on percent difference?*

## SCIENCE FAIR IDEA

The shape of a candy can affect how well many of those candies pack together. Use the water displacement test on a couple differently shaped candies to determine the actual volume of a single candy. Then fill a measuring glass with a certain amount of each type of candy, one type at a time (without water). Divide the volume the candy took up by the number of candies to determine how much space one candy took up, on average, when taking packing into account. *How much space does each type of candy take up in the measuring glass (when packing is taken into account) compared with the actual volume of one candy? In other words, which types of candies pack together the best? How do you think their shape affects this?*

## SCIENCE FAIR IDEA

You can use this same experiment to find the best formula to calculate any other volume. Try using it for an egg, a football, an apple, a bar of soap, other types of candies, or any other irregularly shaped object. Just make sure that you choose an object that can be safely submerged in water! *Which formula is the best?*

## OBSERVATIONS AND RESULTS

Did you find the ellipsoid formula gave the closest answer to the actual volume you measured for one M&M candy?

Using the water displacement test, you should have found the actual volume of a single M&M candy to be about 0.60 to 0.65 cubic centimeter (milliliter). (Adding 100 M&Ms to 100 milliliters of water should have caused the water level to rise to about 160 to 165 milliliters.) When using the sphere formula with the long radius, the calculation gives a volume (about 1.3 cubic centimeters) that is a little bigger than the actual volume of the M&M candy. If you look closely, you will see that the shape of an M&M is not quite perfectly round but is shaped like a sphere that has been squished on one side. If it were not "squished," the sphere formula with the long radius would fit. The cylinder formula also gives a volume (about 1 cubic centimeter) that is too big because it assumes that the entire length of the M&M is as wide as its short diameter is, but it actually tapers around the edges. The sphere formula with the short radius gives a volume (about 0.2 cubic centimeter) that is much smaller than the actual volume of the M&M. The ellipsoid formula should give a volume (about 0.6 cubic centimeter) that is very close to the actual volume of the candy. An M&M indeed has an ellipsoid shape, specifically, a type called an oblate spheroid.

## CLEANUP

If portions of the clay or Play-Doh have been dyed by the M&Ms' coating, you can try to pinch these parts out and throw them in the trash. If you want to eat the soggy M&Ms when you are done with this activity, you may do so, but you should not eat the M&Ms that were in contact with the clay or Play-Doh.

# Statistical Science

## Melt-in-Your-Mouth Math

WHAT ARE THE ODDS YOU'LL SELECT A PARTICULAR COLOR OF M&M WHEN REACHING INTO THE BAG? THIS TASTY ACTIVITY WILL HELP YOU DISCOVER THE ANSWER!

### PROJECT TIME

45 to 60 minutes

### KEY CONCEPTS

Mathematics
Statistics
Probability
Prediction

Have you ever wondered what outcomes are predictable? For example, if you're randomly picking a chocolate out of a box of mixed selections, can you predict how likely it is that you'd end up with a specific type of chocolate, such as a caramel-filled candy? To investigate this tasty topic, in this activity you'll determine the frequency of different color M&Ms in a package of M&M candies. Standard M&Ms come in six different colors: red, green, yellow, blue, orange, and brown. What do you think will be the most, and least, common colors in a typical package of M&Ms?

## BACKGROUND

Often, the frequency of outcomes for an event can be modeled and predicted using statistical analysis. Statistics are facts or data based on known information. The set of information is often about a certain group, and the statistics are a numerical way to describe that group, also called a "population." For example, think about a classroom of students (a population made up of a group of individuals). Some numerical information that could be collected about the population would be the numbers of boys and girls, along with the height of each student. This information could then be used to draw certain conclusions about the population, such as the percentage of students who are female or average height. These facts are statistics about the population.

Many statistics reveal trends, which can be used to form hypotheses and make predictions about certain things. For example, if you figured out the average height of students in a classroom, that statistic could help you predict the average height of students in other classrooms of the same grade at the same school.

## MATERIALS

- At least three standard-size packages of plain M&Ms
- Piece of paper and pencil or pen
- Calculator

## PREPARATION

- Prepare a clean, flat surface on which there is room enough for you to empty the M&Ms packages and make separate piles of each color candy.
- On your piece of paper you might want to create a table to record your results; this will make it easier to do your statistical analysis. If you do this, along the top of the paper make columns labeled "Candy Color," "Package 1," "Package 2," "Package 3," "Total," and "Percentage." (Add more columns if you are using more than three packages.) In the "Candy Color" column make a row for each color (by labeling each row "Brown," "Blue," "Green," "Yellow," "Orange," or "Red"). Fill out the table accordingly as you do the activity.

## PROCEDURE

- Open only the first package of M&Ms and carefully empty it onto the cleaned surface. Count the number of M&Ms of each color. *How many M&Ms are there of each color?* Write the numbers in your record table. (Do not eat any of the M&Ms before you finish the activity!)

- Open each of the other packages of M&Ms separately and again count the number of M&Ms of each color in each package. *How many M&Ms are there of each color in the other packages? Are the numbers similar?* Record the numbers in your table.

- For each candy color, calculate the total number from all of the packages combined. *What is the total of each M&M color for all packages combined?* Write the numbers in your table.

- Next, calculate the total number of all of the M&Ms in all three packages combined. *How many M&Ms are there total?* Write this number somewhere. (It doesn't need to be on a specific place in your table.)

- Lastly, calculate the average percentage of each color M&M. Do this by dividing the total number of each M&M color (for all packages combined) by the total number of M&Ms (in all packages combined). For example, if there are a total of 15 red M&Ms (in all the packages) and 150 M&Ms total (in all packages combined), you would divide 15 by 150 (which equals 0.10, or 10 percent). *What is the average percentage of each candy color?* Write the numbers in your table.

- Look over your results. *In an average package of M&Ms, which color would you predict to be the most common? Which color would you predict to be the least common? Do any of the colors have the same frequencies? Do you see any other trends in your data?*

## EXTRA

*In this activity, you calculated the average number of each candy color, but what were the percentages like for each package you investigated?* You can go back to your data and figure this out. *For the packages you investigated, how much variation was there in the percentage of each color between the different packages?*

## EXTRA

In this activity, you used three packages of M&Ms, but your statistical analysis would probably be more accurate if a larger number of M&Ms packages were used. Try doing this activity with at least five more packages of plain M&Ms. *How do your results change as you use a greater number of packages?*

## SCIENCE FAIR IDEA

You could repeat this activity using almost any other product that comes in a mixture of colors, shapes, sizes, or types. Here are some more ideas to try: Skittles, jelly beans, a bag of chips, marbles, trading cards, etcetera. *Can you accurately predict what you are most likely to pick from one of these packages?*

## OBSERVATIONS AND RESULTS

Were there one or two colors that were consistently the most common in each package of M&Ms? What does this say about the likelihood of randomly selecting one of these colors versus a less common color when reaching into a package of M&Ms?

In statistics, how often a certain event happens is referred to as the frequency of that event. In this activity, the events investigated were different M&M colors and their frequency was measured by counting their numbers within packages of M&Ms (which was the population investigated). Here you most likely found that in a package of plain M&Ms the color with the average highest frequency was blue or orange (in other words, blue or orange was the most common color). These are generally the most common M&M colors in a package of the plain variety of the candies, typically making up around 18 percent to 31 percent of the M&Ms in a package. (Because the percentages of other colors can be close and there is package-to-package variation, it is possible that this trend was not observed when investigating only three packages of M&Ms.) The color with the average lowest frequency (or, in other words, the least common color) was probably brown, red, or yellow. There was likely, however, some variation, and the most (or least) common color may have differed from package to package.

## CLEANUP

When you're all done with your statistical analysis, you may enjoy a tasty M&M snack! And don't forget to share with friends and helpers.

# Suited Science

## What Are the Odds of Drawing That Card?

WHAT IS THE BEST STRATEGY WHEN PLAYING A CARD GAME LIKE "GO FISH"? TRY THIS ACTIVITY TO LEARN HOW YOU CAN BETTER DECIDE WHICH CARDS TO ASK FOR!

Have you ever been playing cards and wished you could use psychic powers to draw the card you wanted? You may not be psychic, but you can still have the power of probability on your side. In this activity, you'll investigate the probabilities of drawing specific types of cards from a deck. You'll discover how math can help you avoid the dreaded phrase, "Go fish!"

### PROJECT TIME

1 to 2 hours

### KEY CONCEPTS

Mathematics
Probability
Chance
Strategy
Cards

## BACKGROUND

When you draw a card from a deck, you have a certain chance of getting a specific type of card, such as a spade or face card, or one particular card, such as the queen of hearts. Consider the game "Go Fish" with a regular card deck. The goal is to get the most four-of-a-kind sets by asking your opponent for matching cards or by drawing them from the deck. To win, you can rely on chance or you can increase your probability of getting matching cards, but how?

By understanding how chance is related to math, you can play with a winning strategy. For example, if you have three kings and one queen in your hand and it's your turn to ask for a card, which one should you ask your opponent for? You might think you should ask for a king, but it's actually better to take a queen! Why? Because you have a better chance of getting it. There are four kings and four queens in the deck, and with three kings and one queen in your hand, there's one king and three queens left. This gives you only one chance to get a king, but three chances to get a queen out of the remaining cards.

## MATERIALS

- Standard deck of playing cards
- Piece of paper
- Pencil or pen
- Calculator

## PREPARATION

- Count the cards in the deck and make sure it is complete. (There should be 54 cards total, including jokers.) Take out the two jokers. Shuffle the deck three times and set aside.
- Pick four types of cards to investigate, such as a color, suit, number, or face card, and a specific one. For example, you could investigate red cards, spades, kings, and the queen of hearts.

- Draw a table to in which to record your data. Make a column for each card type you'll investigate. In the first row, write how many of that type of card are in the deck. For example, there are 26 red cards in a deck, 13 spades, four kings, and one queen of hearts. Make 10 rows below this one for the 10 trials you will be doing.

## PROCEDURE

- Decide which type of card you will investigate first. Draw cards from the top of the deck and flip them over one at a time, counting as you go, and stop when you see that type of card. *How many cards did you draw until you reached that card?* Write down the answer in your table.

- Shuffle the deck again and repeat this process, flipping over the cards and looking for the same type of card. *How many cards did it take this time?* Write down the answer in your table. Repeat this for a total of 10 times for one type of card.

- Repeat this process for each of the four types of cards you picked to investigate. This means that you will have looked for each type of card a total of 10 times.

- Calculate the average number of cards you drew to reach each type of card. Label the last row in your table "Averages" and write them in this row.

- *Which types of cards were the easiest to draw? Which were the most difficult? How do you think the chances of drawing a card relate to the total number of that card type in the deck?*

- *Based on what you saw in this activity, how do you think probability can help you choose the right strategy in a card game?*

## EXTRA

A more advanced way of showing the results of your experiment would be to make histograms, which are a type of graph to show distributions. Try making a separate histogram for each type of card you tested by graphing the number of cards drawn for each trial separately in a bar graph. *When all of the bars are lined up next to each other, what does the overall shape of the distribution look like?*

## SCIENCE FAIR IDEA

The probability of drawing a particular type of card also depends on the number of cards drawn each time. Try doing this activity again but draw samples of three, five, or seven cards at a time. *Do your chances improve as more cards are taken?*

## SCIENCE FAIR IDEA

Probabilities can change your strategies for playing a card game. *Can you design an experiment to show how probabilities can help you choose cards and win "Go Fish"? What about other popular card games? Can you invent your own game based on probabilities?*

## OBSERVATIONS AND RESULTS

Did it take fewer draws to reach a certain color than it took to reach a certain suit or kind of card? Did it take even more draws to reach a specific card?

Mathematicians measure probability by counting and using some very basic math, like addition and division. For example, you can add up the number of spades in a complete deck (13) and divide this by the total number of cards in the deck (52) to get the probability of randomly drawing a spade: 13 in 52, or 25 percent. If you were investigating red cards, kings, or the queen of hearts, the odds of randomly drawing one of these from a complete deck are 50 percent (26 in 52); about 7.7 percent (four in 52); or about 1.9 percent (one in 52), respectively. This is why, on average (when done over enough trials), it is easier to draw a red card than a spade, a spade than a king, and a king than the queen of hearts. As you draw cards from a deck, the odds of finding your card change. For example, if you are looking for a spade and do not get it on your first draw, there are still 13 spades in the deck, but the deck now holds only 51 cards, so your odds of drawing a spade on the second draw are 13 in 51, or about 25.5 percent. This may not seem like much of an improvement, but with every draw the odds continue to increase.

## CLEANUP

Return all materials to where you found them.

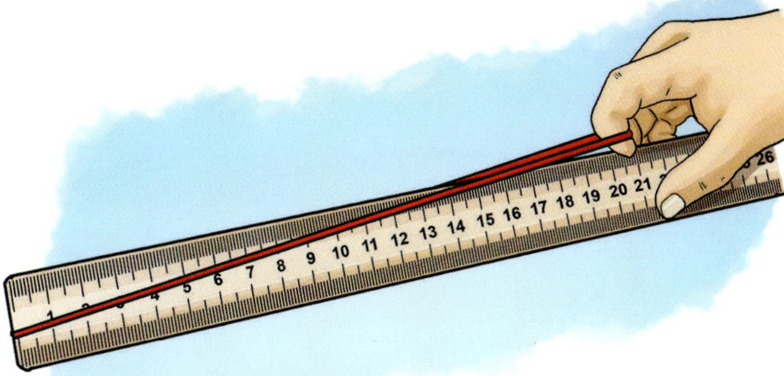

# Snappy Science

## Stretched Rubber Bands Are Loaded with Potential Energy!

DOES THE AMOUNT YOU STRETCH A RUBBER BAND CHANGE HOW FAR IT CAN BE LAUNCHED? GIVE THIS ACTIVITY A TRY TO FIND OUT!

If you've ever been shot with a rubber band then you know it has energy in it—enough energy to smack you in the arm and cause a sting! But have you ever wondered what the relationship is between a stretched rubber band at rest and the energy it holds? The energy the rubber band has stored is related to the distance the rubber band will fly after being released. So can you guess one way to test how much energy a stretched rubber band contains?

### PROJECT TIME

45 to 60 minutes

### KEY CONCEPTS

Physics
Mathematics
Energy
Projectiles

## BACKGROUND

No mechanical contraption would be any fun if it did not work. But "work," in the physics sense, takes energy. Consider a rope and pulley that bring a bucket up a well. The energy that makes this mechanical system work is provided by a person who pulls up the rope.

There are actually two different kinds of energy: potential energy, which is stored energy, and kinetic energy, which is energy in motion. A great example of the difference between kinetic and potential energy is from the classic "snake-in-a-can" prank. This is an old joke where you give someone a can of peanuts and tell them to open it, but inside is actually a long spring that pops out when the lid is twisted off. Because the spring is usually decorated to look like a snake, this prank usually causes the victim to jump back and shout in surprise! When the snaky spring is compressed and secured inside the unopened can, it has potential energy. But when the can is opened, the potential energy quickly converts to kinetic energy as the fake snake jumps out.

## MATERIALS

- A long, wide concrete sidewalk, driveway or other hard surface that you can draw on with chalk (as an alternative, you can make distance markers out of paper and place them on a surface on which you cannot draw)
- Sidewalk chalk
- Metric ruler
- Rubber bands (all of the same length and kind)
- A helper
- Metric tape measure
- Paper and pencil or pen

## PREPARATION

- Find a helper, gather your supplies, and go outside to do this activity. You will want a place with a lot of clearance that has a concrete or other hard surface on which you can draw with chalk.

- Your partner will draw circles around where the flying rubber bands land, so choose a person with a keen eye and some running shoes!
- Use caution to shoot the rubber bands out in front of you, and make sure no one is in the flight path! If necessary, have an adult do the rubber band launching.

## PROCEDURE

- At the outside place you picked, stand where there is lots of clearance in front of you. With your chalk, draw a line in front of your toes. This is where you will line your feet up when you shoot your rubber bands. This is also the mark from where you will measure the distances your rubber bands have flown.
- Your helper can stand a few meters in front of you, but off to the side, not directly in the line of fire! Make sure he or she has a piece of chalk.
- Shoot a rubber band by hooking it on the front edge of the ruler, then stretching it back to 10 centimeters on the ruler and letting the rubber band go. Remember the angle and height at which you hold the ruler because you will need to keep it the same for each rubber band launch.
- Have your helper draw a small chalk circle where the rubber band landed.
- Shoot at least four more rubber bands in the same way, stretching them back to 10 centimeters on the ruler each time. Have your helper circle where each lands.
- Measure the distances from your line to the circles your helper made. Write these distances down under the heading "10 centimeters." *Did all five rubber bands land close to each other or was there a lot of variation in where they fell?*

- Shoot more rubber bands in the same way, except stretch them back to 15 centimeters, 20 centimeters, 25 centimeters, or 30 centimeters. Shoot at least five rubber bands for each stretch length. After each launch, have your helper circle where they land. After launching five rubber bands at a given stretch length, measure the distances from your line to the circles. Write these distances under a heading for their stretch length (for example, "20 centimeters").
- *For each stretch length, did all five rubber bands land close to one another or was there a lot of variation? Did they land far from where the rubber bands landed that were launched using different stretch lengths?*
- Average your results for each stretch length and make a graph of your results by putting "Stretch Length (cm)" on the x-axis (this will be 10 cm, 15 cm, 20 cm, 25 cm, and 30 cm) and "Launch Distance (cm)" on the y-axis (this will be the distances you measured). *Do your data follow any type of pattern or trend? What was the relationship between the stretch length and the launch distance? What do you think this indicates about the relationship between potential and kinetic energy when using rubber bands?*
- **Tip:** If you run out of rubber bands, you can always grab some of the ones you already used and reuse them because there will be a chalk circle where they landed.

## EXTRA

In this activity you kept the angle and height of the launch the same from trial to trial. *How do these variables affect the distance the rubber band travels?* Design a separate activity to test each of these variables separately.

## SCIENCE FAIR IDEA

You can do a very similar activity to this one by using other types of mechanical systems, such as springs and slingshots. *How do the data collected using these other mechanical systems compare with that collected using rubber bands?*

## EXTRA

For an advanced challenge, you can use linear regression to further analyze your data. *Can you define an equation that expresses the relationship between potential and kinetic energy in this system?*

## OBSERVATIONS AND RESULTS

Did the rubber bands stretched to 30 centimeters launch farther than the other rubber bands? Did you see a linear relationship between the launch distance and stretch length when you graphed your data?

You input potential (stored) energy into the rubber band system when you stretched the rubber band back. Because it is an elastic system, this kind of potential energy is specifically called elastic potential energy. Elastic potential energy (measured in the unit joules) is equal to 1/2 multiplied by the stretch length ("x") squared, multiplied by the spring constant "k." The spring constant is different for every rubber band, but can be figured out. When the rubber band is released, the potential energy is quickly converted to kinetic (motion) energy. This is equal to one-half the mass (of the rubber band) multiplied by its velocity (in meters per second) squared.

Using these equations, you can calculate the velocity of the rubber band right when it is released, and find that the velocity has a linear relationship with the stretch length. (Because the amount of time that the rubber band spends in the air is dependent on its initial height and force of gravity, and these factors should not change between your trials, then how far the rubber band flies depends on its initial velocity.) Consequently, after you graph your data, you should see a roughly linear relationship between the stretch length and the launch distance.

## CLEANUP

Return all materials to where you found them.

# Probability and the Birthday Paradox

HOW MANY PEOPLE WOULD YOU HAVE TO ASK UNTIL YOU FOUND SOMEONE WITH THE SAME BIRTHDAY AS YOU? THE ANSWER MIGHT SURPRISE YOU!

## PROJECT TIME

30 minutes to 1 day

## KEY CONCEPTS

Mathematics
Probability
Statistics

Have you ever noticed how sometimes what seems logical turns out to be proved false with a little math? For instance, how many people do you think it would take to survey, on average, to find two people who share the same birthday? Due to probability, sometimes an event is more likely to occur than we believe it to. In this case, if you survey a random group of just 23 people, there is actually about a 50-50 chance that two of them will have the same birthday. This is known as the birthday paradox. Don't believe it's true? You can test it and see mathematical probability in action!

## BACKGROUND

The birthday paradox, also known as the birthday problem, states that in a random group of 23 people, there is about a 50 percent chance that two people have the same birthday. Is this really true? There are multiple reasons why this seems like a paradox. One is that when in a room with 22 other people, if a person compares his or her birthday with the birthdays of the other people it would make for only 22 comparisons—only 22 chances for people to share the same birthday.

But when all 23 birthdays are compared against each other, it makes for much more than 22 comparisons. How much more? Well, the first person has 22 comparisons to make, but the second person was already compared to the first person, so there are only 21 comparisons to make. The third person then has 20 comparisons, the fourth person has 19 and so on. If you add up all possible comparisons (22 + 21 + 20 + 19 + ... +1) the sum is 253 comparisons, or combinations. Consequently, each group of 23 people involves 253 comparisons, or 253 chances for matching birthdays.

## MATERIALS

- Groups of 23 or more people (10 to 12 such groups) or a source with random birthdays (see Preparation below for tips)
- Paper and pen or pencil
- Calculator (optional)

## PREPARATION

- Collect birthdays for random groups of 23 or more people. Ideally you should get 10 to 12 groups of 23 or more people so you have enough different groups to compare. (You don't need the year for the birthdays, just the month and day.)
- **Tip:** Here are a few ways that you can find a number of randomly grouped people: Ask school teachers to pass a list around each of their classes to collect the birthdays for students in the class (most schools have around 25 students in a class); use the birthdays of players on major league baseball teams (this information can easily be found on the internet); or use the birthdays of other random people using online sources.

## PROCEDURE

- For each group of 23 or more birthdays that you collected, sort through them to see if there are any birthday matches in each group.
- *How many of your groups have two or more people with the same birthday? Based on the birthday paradox, how many groups would you expect to find that have two people with the same birthday? Does the birthday paradox hold true?*

### EXTRA

In this activity you used a group of 23 or more people, but you could try it using bigger groups. If you use a group of 366 people—the greatest number of days a year can have—the odds that two people have the same birthday are 100 percent (excluding February 29 leap year birthdays). *What do you think the odds are in a group of 60 or 75 people?*

### EXTRA

Rolling dice is a great way to investigate probability. You could try rolling three 10-sided dice and five six-sided dice 100 times each and record the results of each roll. Calculate the mathematical probability of getting a sum higher than 18 for each combination of dice when rolling them 100 times. *Which combination has a higher mathematical probability, and was this true when you rolled them?*

## OBSERVATIONS AND RESULTS

Did about 50 percent of the groups of 23 or more people include at least two people with the same birthdays?

When comparing probabilities with birthdays, it can be easier to look at the probability that people do not share a birthday. A person's birthday is one out of 365 possibilities (excluding February 29 birthdays). The probability that a person does not have the same birthday as another person is 364 divided by 365 because there are 364 days that are not a person's birthday. This means that any two people have a 364/365, or 99.726027 percent, chance of not matching birthdays.

As mentioned before, in a group of 23 people, there are 253 comparisons, or combinations, that can be made. So, we're not looking at just one comparison, but at 253 comparisons. Every one of the 253 combinations has the same odds, 99.726027 percent, of not being a match. If you multiply 99.726027 percent by 99.726027 253 times, or calculate (364/365)253, you'll find there's a 49.952 percent chance that all 253 comparisons contain no matches. Consequently, the odds that there is a birthday match in those 253 comparisons is 1 - 49.952 percent = 50.048 percent, or just over half! The more trials you run, the closer the actual probability should approach 50 percent.

## CLEANUP

Return any materials to where you found them.

# Juice Box Geometry

ARE ALL JUICE BOXES CREATED EQUALLY? THIS ACTIVITY WILL HELP YOU FIND OUT!

## PROJECT TIME

1 to 2 hours

## KEY CONCEPTS

Mathematics
Geometry
Psychology

Juice boxes sure are convenient—just poke a straw in and sip away! But have you ever noticed that some juice boxes don't seem to have as much juice as others, even when they have a lot of packaging? You might be amazed how much thought goes into designing a juice box. Each manufacturer has carefully calculated how big each side of the box needs to be to hold a certain amount of juice inside and how altering the box's dimensions affects its overall appearance and packaging. Investigating the geometry of juice boxes can reveal how some have more efficient packaging than others. And how appearances can be deceptive!

## BACKGROUND

Geometry uses math to describe and investigate different points, lines, and shapes. In geometry, the relationships among different parts of a shape can be described with mathematical formulas. Formulas can be used to describe the relationship among the three dimensions of real-world objects: length, width, and height. And that information can help us learn about objects around us.

A rectangular box, such as a juice box or cereal box, is called a rectangular prism in geometry-speak. If you measure the height, length, and width of a rectangular prism, you can use that information in a formula to calculate the volume and surface area of that object. The volume is the amount of space an object fills in three-dimensional space (which tells approximately how much a container, like a juice box, can hold), and the surface area is the total amount of area on the outer surface of the object (which tells approximately how much material was used to create the shape).

## MATERIALS

- Metric ruler
- Juice boxes, five or more different brands (one box per brand)
- Calculator
- Paper
- Pen or pencil

## PREPARATION

- Using a ruler, for each juice box measure its length (how long it is from left to right), width (how deep it is from front to back), and height. Write down these measurements for each box (in centimeters). *Are the dimensions fairly similar between the different boxes, or is there a lot of variation?*

## PROCEDURE

- For each juice box, calculate the surface area and write it down. The surface area of a rectangular prism = 2 X length X height + 2 X width X height + 2 X length X width. (The units will be in $cm^2$, or centimeters squared.) *How much do the surface areas vary for the different boxes?*

- For each juice box, calculate the volume and write it down. The volume of a rectangular prism = length X height X width. (The units will be in $cm^3$, or centimeters cubed.) *How much do the calculated volumes vary for the different boxes?*

- Compare the volume of juice in each box with its calculated volume. The volume of juice in the box is usually found on the front of the juice box near the bottom. Use the volume listed in milliliters (ml) because 1 ml equals 1 $cm^3$. *How close is your calculated volume to the actual volume of juice in each box?*

- For each juice box, calculate the ratio of surface area to volume. You can do this by dividing its calculated surface area by the volume of juice it holds (again in milliliters). If the answer of the calculation is close to 1, the surface area and volume are almost equal. The higher the number, the more packaging is used per volume of juice; the lower the number, the less packaging is used per volume of juice. *Which box uses the least amount of packaging for the volume of juice it holds? Which box uses the most?*

- *Overall, which juice box is the most efficient at holding juice for its size? Which one has a juice volume closest to its calculated volume and has the smallest surface-area-to-volume ratio?*

## EXTRA

Packaging comes in all shapes and forms. Try to use the principles of geometry to make geometric models to compare the efficiency of other shapes of packaging. *For example, how do cylindrical ice cream tubs compare with rectangular ice cream tubs?*

## EXTRA

Cereal containers mostly come in rectangular shapes but can vary greatly in size. You can use different cereal boxes to investigate how geometry can help reduce waste in packaging material. *Is there less wasted packaging in a large box or a small box of cereal? Is it better to buy single-use cereal boxes that are only one portion size or large, multi-portion size boxes?*

## OBSERVATIONS AND RESULTS

Was the actual volume of juice in some juice boxes much closer to the calculated volume than when compared with other boxes? Did some boxes have a much smaller surface-area-to-volume ratio than others?

As you probably saw from this activity, different juice boxes that hold the same amount of juice can have very different dimensions. This can change their surface areas and the total volumes that they can hold. The actual volume of juice held by many juice boxes takes up over 95 percent of the total calculated volume of the box, although some boxes may hold juice that takes up less of the total calculated volume than this. The surface-area-to-juice-volume ratio is usually a little over 1, often around 1.2 to 1.3. This means that there is a little more packaging used per volume of juice. The higher the ratio, the more packaging is used per volume of juice.

## CLEANUP

Save the juice boxes to drink later. Return all other materials to where you found them.

# The Tower of Hanoi

TRICKY TOWERS: SEE HOW MATH CAN HELP SOLVE THIS AGE-OLD PATTERN PUZZLE.

Are you tired of math worksheets and homework? Did you know that there are more creative ways to exercise your mathematical muscle? A lot of games, puzzles, and riddles revolve around mathematical concepts. Think about simple games such as tic-tac-toe, more strategic games such as chess, or math puzzles such as sudoku. People have been playing these games and puzzles for centuries! They are fun, entertaining, and sometimes useful. See how you enjoy this one.

## PROJECT TIME

2 to 3 hours

## KEY CONCEPTS

Mathematics
Patterns
Algorithms
Puzzles

## BACKGROUND

Throughout history, mathematics has fulfilled many practical needs such as measuring plots of land, studying astronomy, and calculating taxes. But math can also be used for entertainment—mathematical games, riddles, challenges, and puzzles are also interwoven throughout history!

The Tower of Hanoi (also called the Tower of Brahma or the Lucas Tower) was invented by a French mathematician Édouard Lucas in the 19th century. It is associated with a legend of a Hindu temple where the puzzle was supposedly used to increase the mental discipline of young priests. In the legend, the young priests were given 64 gold discs stacked neatly on one of three posts. Each disc rested on a slightly larger disc. The priests' goal was to re-create the stack on a different post by moving discs, one at a time, to another post with the rule that a larger disc could never be placed on top of a smaller disc. Using mathematics, you can calculate that even when the priests found the most efficient way to solve the problem and moved the discs at a rate of one per second, it would take almost 585 billion years to finish the job. That is more than 40 times the age of the universe!

You might wonder how mathematics is involved in playing this game. As you play the game with more and more discs, you will notice you start to look for patterns. If you try to explain how you solve the puzzle, you might realize you use one of the following mathematical concepts:

—Iterative solutions, where the same sequence of instructions is repeated over and over
—Recursive solutions, where you use information from one step to find the next step
—Patterns and translating these in mathematical formulas.

These terms might seem hard to understand. Don't worry! Play the game, and you might realize that you develop these "solutions" on your own.

## MATERIALS

- Five different-size discs such as buttons, bottle caps, container lids, etcetera (The biggest disc needs to have a diameter no larger than 5 inches [12.7 cm].)
- Cardboard, about 5 by 15 inches (12.7 by 38 cm)
- Marker
- Ruler

## PREPARATION

- Draw two straight lines to divide the cardboard into three equal-size squares (about 5 by 5 inches [12.7 by 12.7 cm] each).

## PROCEDURE

- Start the game with your two smallest discs. Stack them on the leftmost square of your cardboard, with the smaller disc on top of the larger disc.
- The starting position of the game is a tower on the leftmost square of the board (like the two-disc tower you have now). The goal of the game is to move the tower to the rightmost square of the board while following these rules:
  - You can only move one disc at a time.
  - You can only move the topmost disc from any square.
  - Discs can be placed on an empty square on the board or on a larger disc, but discs cannot be placed next to each other on top of a larger disc.
- *Looking at the rules, can you see it is not allowed to place a disc off the board? Where does it state you cannot start two towers on one square?*
- Try the game with your two-disc tower. *Can you move the tower to the rightmost square, following the rules above? How many moves did you need?* If you get confused or cannot finish the game, ask a friend to help you.
- Replay the game with the two-disc tower. *Do you think you can find the "best" solution, with the smallest possible number of moves?*
- Once you master the two-disc tower, try with three discs. Stack three discs (from largest to smallest) on the leftmost square and start over. *How many moves did you need this time?*

- Try the game again with three discs. *Can you finish the game with fewer moves? Do you think you can finish the game with the smallest possible number of moves?*
- Now try with four discs. *Can you find a way to do it in as few moves as possible? Does anything you learned while solving the three-disc tower help you solve the four-disc one?*
- Play the game a few more times and observe closely. *Do you have a strategy? Could you explain to a friend how you finish the game?*
- Challenge yourself and try the five-disc tower. *If you had a strategy, can you extend it to the five-disc tower?*
- If you did not have a strategy yet, do not give up! See if you can find a pattern now. *Can you find the smallest number of moves needed to finish the five-disc game? How many more moves would you need compared with the four-disc one? How many more moves are needed to move the four-disc tower compared with the three-disc one?*

## EXTRA

*Did you find a single strategy that works to solve the puzzle for two, three, four, or five discs? Can you write down the steps for this strategy so you always remember how you solved the puzzle?* This list of steps is called an "algorithm."

## EXTRA

*Can you try your algorithm on a different number of discs? Does it still work, or can you adjust it to make it work?*

## EXTRA

*If you found the minimal number of moves required to solve the puzzle for two, three, four, or five discs, can you see a pattern in the minimal number of moves required? Could you use the pattern to predict the minimal number of moves required for a game with six discs?* Maybe you can even find an equation to calculate the minimum number of moves required to play a game with a certain number of discs. (Note: the equation involves exponents.)

## EXTRA

There are many variations of the Tower of Hanoi game. For example, you can arrange the squares in a circle and only allow discs to move clockwise. Look up some other variations of the game. *How do the new rules change the solutions you can find? Can you write algorithms to solve the modified versions?*

# OBSERVATIONS AND RESULTS

Were you able to move the two-disc stack in three moves? Three is the minimal number of moves needed to move this tower. Maybe you also found in the games three discs can be finished in seven moves, four discs in 15, and five discs in 31.

To play the two-disc game with only three moves, place the small disc on the middle square, the bigger one on the rightmost square and finish by placing the smaller disc on the bigger one. A way to play the three-disc game is to move the topmost two-disc stack to the middle square. This is a slight variation of the two-disc game and can be done in three moves. Then move the biggest disc from the leftmost to the rightmost square (one move). Next, move the two-disc stack that is now in middle square on top of the biggest disc. Again, a slight variation of the two-disc game, which can be done with three moves. This yields 3 + 1 + 3, or seven moves. A similar approach helps play the four-disc game. First, move the topmost three-disc stack to the middle square using seven moves. Then move the biggest disc from the leftmost to the rightmost square (one move). Now move the three-disc stack that is in middle square on top of the biggest disc (seven moves). This yields exactly 7 + 1 + 7, or 15 moves. Mathematicians call this is a recursive procedure. To solve the puzzle with more discs, you start out with the solution for a smaller number of discs.

There are other ways to play the game. There is a pattern used to start the game when played with a minimal number of moves: The smallest disc goes to the rightmost square when playing with an even number of discs and to the middle square when playing with an odd number of discs. Maybe you also noticed you alternate between moving the smallest and the next-smallest movable disc while playing the game. If you pay close attention, you might note that you repeat the same set of instructions over and over until the tower appears on the other side of your board. Mathematicians call this is an iterative solution.

## CLEANUP

Return all materials to where you found them.

# THE SCIENTIFIC METHOD

The scientific method helps scientists—and students—gather facts to prove whether an idea is true. Using this method, scientists come up with ideas and then test those ideas by observing facts and drawing conclusions. You can use the scientific method to develop and test your own ideas!

Question: What do you want to learn? What problem needs to be solved? Be as specific as possible.
Research: Learn more about your topic and refine your question.
Hypothesis: Form an educated guess about what you think will answer your question. This allows you to make a prediction you can test.
Experiment: Create a test to learn if your hypothesis is correct. Limit the number of variables, or elements of the experiment that could change.
Analysis: Record your observations about the progress and results of your experiment. Then analyze your data to understand what it means.
Conclusion: Review all your data. Did the results of the experiment match the prediction? If so, your hypothesis was correct. If not, your hypothesis may need to be changed.

# GLOSSARY

**atom:** The basic building blocks for all matter.
**chance**: The likelihood something in particular will happen.
**convert:** To change into another form.
**energy**: The ability to do work.
**height:** The distance from the top of an object to the bottom.
**length:** The distance from one end of an object to the other end.
**meniscus:** A curve in the surface of water when it touches something else.
**odds**: The probability that one thing will happen instead of another.
**paradox:** Something that seems to go against itself but really may be true.
**population:** A set of items with at least one thing in common.
**prism:** A see-through object that has several flat surfaces.
**radius:** The distance from the middle of a circle to any point on the edge of the circle.
**ratio:** A way to compare two things that are the same type.
**velocity:** A measure of how fast something moves in a specific direction.
**width:** The distance of the side of an object.

# ADDITIONAL RESOURCES

## Books


Cheng, Eugenia. *Molly and the Mathematical Mysteries: Ten Interactive Adventures in Mathematical Wonderland*. Somerville, MA: Big Picture Press, 2020.

Davis, Ben Ffrancon and Mubeen, Junaid. *What's the Point of Math?* New York, NY: DK Publishing, 2020.

Woo, Eddie. *It's a Numberful World: How Math is Hiding Everywhere*. New York, NY: The Experiment, LLC, 2019.


## Websites

**Discovery Education**
*sciencefaircentral.com*

**Exploratorium**
*www.exploratorium.edu/search/science%20fair%20projects*

**Science Buddies**
*www.sciencebuddies.org/science-fair-projects/project-ideas/list*

**Science Fun for Everyone!**
*www.sciencefun.org/?s=science+fair*

## Videos

**The Art of Math**
*https://wqed.pbslearningmedia.org/resource/mgbh.math.rp.artmath/the-art-of-math/*

**Measure: The Metric System Explained**
*https://wqed.pbslearningmedia.org/resource/measure-metric-system-explained-video-gallery/science-trek/*

**Publisher's note to educators and parents:** Our editors have carefully reviewed these websites to ensure that they are suitable for students. Many websites change frequently, however, and we cannot guarantee that a site's future contents will continue to meet our high standards of quality and educational value. Be advised that students should be closely supervised whenever they access the internet.

# INDEX